Yale Mathematical Monographs 7

Editorial Committee

Nathan Jacobson

Shizuo Kakutani

William S. Massey

George D. Mostow

James K. Whittemore Lectures in Mathematics

given at Yale University

The Finite Simple Groups and Their Classification

Michael Aschbacher

Yale University Press

New Haven and London

1980

Published with assistance from the Kingsley Trust
Association Publication Fund established by the
Scroll and Key Society of Yale College.

Copyright © 1980 by Yale University.
All rights reserved. This book may not be
reproduced, in whole or in part, in any form
(beyond that copying permitted by Sections 107
and 108 of the U.S. Copyright Law and except by
reviewers for the public press), without written
permission from the publishers.

Printed in the United States of America.

Published in Great Britain, Europe, Africa, and
Asia (except Japan) by Yale University Press,
Ltd., London. Distributed in Australia and
New Zealand by Book & Film Services, Artarmon,
N.S.W., Australia; and in Japan by Harper & Row,
Publishers, Tokyo office.

Library of Congress Cataloging in Publication Data
Aschbacher, Michael, 1944-
   The finite simple groups and their classification.

   (James K. Whittemore lectures in mathematics given
at Yale University) (Yale mathematical monographs; 7)
   Includes bibliographical references.
   1. Finite simple groups. I. Title.
II. Series: Yale University. James K. Whittemore
lectures in mathematics given at Yale University.
III. Series: Yale mathematical monographs; 7.
QA171.A82   512'.22   79-20927
ISBN 0-300-02449-5

Contents

	Introduction	vii
I.	The Known Simple Groups	1
II.	The Local Theory	16
III.	Fischer's Theory of Finite Groups	28
IV.	Signalizer Functors and Disconnected Groups	35
V.	Groups of Component Type	41
VI.	Groups of Even Characteristic	48
VII.	The Future	55
	References	59

Introduction

During the last decade a program to classify the finite simple groups has begun to take shape. This monograph is a description of that program intended for a general mathematical audience. The material is taken from a series of four Whittemore Lectures given at Yale University in September of 1978.

While a variety of mathematical theories make important contributions to the classification program, perhaps the most significant contribution comes from the <u>local theory of finite groups</u>. For the most part, this theory has not found application in other areas of mathematics and as a result is unfamiliar to most mathematicians. Much of this monograph is concerned with the local theory. On the other hand several parts of the program are not discussed, or discussed only briefly. Most notably there is no discussion of the Odd Order Paper or the classification of the small rank groups which took place in the sixties and which depended heavily on the theory of group characters and upon Brauer's modular theory. While the theory of groups of Lie type is mentioned in Chapter I, more time could have been devoted to that subject. I have however attempted to draw several analogies between algebraic groups and finite groups in the hope that this will be of assistance to those with some background in algebraic or Lie groups. In Chapter III there is a discussion of a theory of finite groups initiated by Bernd Fischer which also has been of importance to the classification program. If anything, this theory is even less well known than the local theory. It is my impression that it is even unfamiliar to most of the specialists.

As I just mentioned, certain small finite groups were classified in the sixties using techniques somewhat different from those most in use today. This is one example of a fairly common mathematical phenomenon. Namely, in analyzing some class of objects one may find it necessary to use one collection of techniques to study most of the members of the class (the generic objects) and a different set of techniques to deal with the remaining members of the class (the small objects). This dichotomy pervades the study of the finite simple groups. In particular, there is a reasonably elegant theory which can be used to handle the generic finite simple groups, while the small groups appear to require intricate ad hoc arguments. I have chosen to discuss only the generic situation, but the reader should be aware that much of the work involved in classifying the simple groups is expended in dealing with the small groups.

This dichotomy also appears to be one of the keys to understanding the finite simple groups and the classification program. It seems to be the case that there are an infinite family of generic simple groups in the groups of Lie type, plus a small number of exceptional simple groups in the sporadic groups. There is no reasonable method known for describing all these groups at the same time, and I suspect that no such method will appear. It is natural that this state of affairs should be reflected in the mechanics of the proof.

The monograph is organized as follows: Chapter I consists of a discussion of the known finite simple groups. Chapter II is an introduction to the local theory and to the basics of the classification program; Chapter III discusses Fischer's theory of finite groups. The next three chapters outline the major steps in the classification program. The last chapter summarizes those steps in the program which remain to be completed and speculates about the postclassification universe.

The standard elementary reference in the area of finite simple groups is Gorenstein's text Finite Groups [20]. I assume familiarity with only the most elementary concepts from that text, concepts which are standard in all areas of group theory. In a few places I have pointed out other texts as references in certain subareas. However, most literature on finite simple groups is confined to the journals. Indeed much of the most recent work has not yet been published, or, in some cases, even been distributed in preprint form. Thus many results are cited without any reference.

I. The Known Simple Groups

The known nonabelian simple groups may be denoted as follows:

                Groups of Lie type

                Alternating groups

                Sporadic groups .

The groups of Lie type are the finite analogues of the complex semisimple Lie groups. Let X_n be a simple Lie algebra over the complex numbers $\mathbb{C}$. n is the rank of the algebra and $X = A, B, C, D, E, F,$ or G is its type. Chevalley has shown that X_n possesses a basis with respect to which certain structure constants are integers, and he produces certain Lie algebra automorphisms leaving invariant the integral span of this basis. Then, given a field K, these facts are used to produce a Lie algebra over K and a group $X_n(K)$ of Lie algebra automorphisms. The groups $X_n(K)$ are the (ordinary) <u>Chevalley groups</u> over K.

The Chevalley group $X_n(\mathbb{C})$ is a semisimple Lie group. If $\cdot$ K is algebraically closed then $X_n(K)$ is a semisimple algebraic group. If $K = GF(q)$ is the finite field of order q then $X_n(K)$ is denoted by $X_n(q)$ and is a finite group which is essentially simple.

By considering the fixed points of certain outer automorphisms of the finite Chevalley groups, one obtains more finite simple groups known as the <u>twisted Chevalley groups</u>. $^r X_n(q)$ denotes the fixed points of an automorphism of $X_n(q^e)$ of order r for a suitable extension field $GF(q^e)$ of $GF(q)$. The Chevalley groups of ordinary and twisted type are the <u>groups of Lie type</u>. There is a complete list of these groups in Table 1.

The groups $A_n(q)$, $^2A_n(q)$, $B_n(q)$, $C_n(q)$, $D_n(q)$, and $^2D_n(q)$ are often referred to as the <u>classical groups</u>. For each can be represented as the group of projective transformations of determinant 1 of a finite dimensional vector space over a finite field, or of such a space with some nondegenerate bilinear, sesquilinear, or quadratic form. In this aspect, the classical groups have been known since the nineteenth century.

These representations give rise to a second notational system for the classical groups, where $L_n(q)$, $U_n(q)$, $PSp_n(q)$, and $\Omega_n^\varepsilon(q)$ denote the n-dimensional projective special linear, unitary, symplectic, and orthogonal group, respectively, and with ε the Witt index of the associated orthogonal space in the case of the orthogonal groups of even dimension. The correspondence between the two notations is indicated in Table 1.

Table 1

The Simple Groups of Lie Type

Lie Notation	Classical Notation
$A_n(q)$, $n \geq 1$	$L_{n+1}(q)$
$B_n(q)$ $n \geq 3$	$\Omega_{2n+1}(q)$
$C_n(q)$ $n \geq 2$	$PSp_{2n}(q)$
$D_n(q)$ $n \geq 4$	$\Omega_{2n}^{+}(q)$
$E_n(q)$, $6 \leq n \leq 8$	
$F_4(q)$	
$G_2(q)$	
$^2A_n(q)$, $n \geq 2$	$U_{n+1}(q)$
$^2D_n(q)$, $n \geq 4$	$\Omega_{2n}^{-}(q)$
$^3D_4(q)$	
$^2E_6(q)$	
$^2B_2(2^{2m+1})$, $m \geq 1$	$Sz(2^{2m+1})$
$^2F_4(2^{2m+1})$, $m \geq 0$	
$^2G_2(3^{2m+1})$, $m \geq 1$	

4 / The Known Simple Groups

The groups $^2B_2(2^n)$ were first discovered by M. Suzuki while proving a classification theorem for finite groups. Only later were these groups observed to be of Lie type. Hence they are often denoted by $Sz(2^n)$.

J. Tits has developed a system of axioms satisfied by a large, interesting class of groups which include the groups of Lie type. A group G is said to have a (B,N)-pair if G has subgroups B and N such that:

(BN.1) $B \cap N = H$ is a normal subgroup of N, and $W = N/H$ is a Coxeter group with fundamental reflections Hu_i, $1 \leq i \leq n$.

(BN.2) G is generated by B and N

(BN.3) For each $u \in N$ and $1 \leq i \leq n$,

$$Bu\, Bu_i \subseteq (BuB) \cup (Buu_iB)$$

(BN.4) u_i does not normalize B, $1 \leq i \leq n$.

A group W is a <u>Coxeter group</u> with <u>fundamental reflections</u> w_i, $1 \leq i \leq n$ if W is the group generated by the elements w_i subject to defining relations

$$|w_iw_j| = m_{ij}, \quad 1 \leq i \leq j \leq n$$

with $m_{ii} = 2$ for each i. Coxeter has obtained a classification of the Coxeter groups.

B is the <u>Borel subgroup</u> of G and W the <u>Weyl group</u> of G. If G is of Lie type over a field of characteristic p then B is the normalizer of a Sylow p-subgroup U of G. Hence G has a <u>split</u> (B,N)-pair. That is G satisfies the extra axiom:

(BN.5) B has a normal subgroup U with $B = UH$ and $U \cap H = 1$.

The number n of fundamental reflections generating the Weyl group is the <u>rank</u> of the (B,N)-pair. If $G = X_n(q)$ is an ordinary Chevalley group then the BN-rank of G is the same as the rank n of the Lie algebra X_n determining G. In any case we refer to the BN-rank of a group of Lie type as its <u>Lie rank</u>.

As a corollary to a much more general result, Tits has shown that any finite simple group with a (B,N)-pair of rank at least 3 is a group of Lie type [39]. G. Seitz and P. Fong have established the result for finite simple groups with a split (B,N)-pair of rank 2, [15].

Notice that G has a (B,N)-pair of rank 1 exactly when G acts as a doubly transitive permutation group on the cosets of B. As there are many doubly transitive groups which are not of Lie type, it is evident that for small ranks, the (B,N)-pair axioms lose some of their strength. Here is our first example of the general principle mentioned in the introduction that small groups require different treatment than the generic group. Hence the Lie rank of a group of Lie type is one good measure of its size. The order of the associated field is another significant measure. Groups of small rank over fields of order 2, 3, and 4 present difficulties at various points in the classification theory.

Despite such difficulties, Hering, Kantor, and Seitz have shown in [25] that a finite simple group with a split (B,N)-pair of rank 1 is either a group of Lie type or a group with the same order and subgroup structure as a group $^2G_2(3^n)$. The groups $^2G_2(3^n)$ are commonly called <u>Ree groups</u> after R. Ree, who discovered them, and groups with the same order and subgroup structure as Ree groups are said to be of <u>Ree type.</u> It seems to be a very difficult problem to show that groups of Ree type are actually Ree groups, although this is almost certainly the case.

The groups of Lie type in characteristic p may also be realized as the fixed points of certain endomorphisms of Chevalley groups over the algebraic closure of GF(p). The theory of algebraic groups may then be invoked to investigate the finite group. This is an extremely profitable point of view.

For more detailed information about the groups of Lie type, the reader is directed to Carter's book on the subject [12]. The book also contains a large bibliography.

The alternating group of degree n is the group of all even permutations on a set of order n. The alternating group of degree n is commonly denoted by A_n, which should not be confused with the group $A_n(q)$ of Lie type. A_n is a subgroup of index 2 in the symmetric group S_n, the group of all permutations on set of order n. S_n is the Weyl group of the group $A_{n-1}(q)$, so in a certain sense the alternating groups are covered by the theory of groups of Lie type. In any case the alternating group of degree n can be effectively studied from the point of view of its permutation representation of degree n.

Nonabelian finite simple groups other than the groups of Lie type and the alternating groups belong to no known family of simple groups. Such groups are therefore termed <u>sporadic groups</u>. To begin this discussion of the sporadic groups, I think it best to make reasonably precise the meaning of the following three concepts as applied to the sporadic groups:

> Discovery
>
> Existence
>
> Uniqueness

I understand a sporadic group to be <u>discovered</u> when a sufficient

amount of self consistent information about the group is available. Such
information typically includes the order, conjugacy classes, subgroup
structure, and character table of the group. To be more precise, let
P be a set of group theoretical properties. One begins by studying
simple groups satisfying P. The sporadic group satisfying P is
<u>discovered</u> when there is a proof that if G is a finite simple group
satisfying P then the information I(G) is determined and self
consistent, where I(G) is the information indicated above.

Notice that under this definition the group can be discovered
before it is shown to exist and that <u>the</u> group discovered could
conceivably be a number of groups with the same properties. Of course the
group is said to <u>exist</u> when there is a proof that there exists some
finite simple group satisfying P, and the group is <u>unique</u> when there
is a proof that if G and H are finite simple groups satisfying P
then G is isomorphic to H.

At present 26 sporadic groups are discovered. These groups are
listed in Table 2. 24 of these groups are known to exist. There is no
existence proof for the groups denoted by F_1 and J_4 in Table 2.
The uniqueness question is not completely settled for F_1 and J_4.

The person who first postulates the existence of the group usually
has his name attached to the group. Sometimes the names of those
involved in the construction of the group are also incorporated in the
group name.

The sporadic groups were discovered in a variety of ways. Mathieu
discovered the first five sporadic groups in the nineteenth century as
multiply transitive permutation groups. A permutation group on a set
Ω is a subgroup of the symmetric group on Ω. Recall that the symmetric

Table 2

The Sporadic Simple Groups

Notation	Name	Order
M_{11}	Mathieu	$2^4 \cdot 3^2 \cdot 5 \cdot 11$
M_{12}		$2^6 \cdot 3^3 \cdot 5 \cdot 11$
M_{22}		$2^7 \cdot 3^2 \cdot 5 \cdot 7 \cdot 11$
M_{23}		$2^7 \cdot 3^2 \cdot 5 \cdot 7 \cdot 11 \cdot 23$
M_{24}		$2^{10} \cdot 3^3 \cdot 5 \cdot 7 \cdot 11 \cdot 23$
J_1	Janko	$2^3 \cdot 3 \cdot 5 \cdot 7 \cdot 11 \cdot 19$
J_2 = HJ	Hall-Janko	$2^7 \cdot 3^3 \cdot 5^2 \cdot 7$
J_3 = HJM	Higman-Janko-McKay	$2^7 \cdot 3^5 \cdot 5 \cdot 17 \cdot 19$
J_4	Janko	$2^{21} \cdot 3^3 \cdot 5 \cdot 7 \cdot 11^3 \cdot 23 \cdot 29 \cdot 31 \cdot 37 \cdot 43$
HS	Higman-Sims	$2^9 \cdot 3^2 \cdot 5^3 \cdot 7 \cdot 11$
Mc	McLaughlin	$2^7 \cdot 3^6 \cdot 5^3 \cdot 7 \cdot 11$
Sz	Suzuki	$2^{13} \cdot 3^7 \cdot 5^2 \cdot 7 \cdot 11 \cdot 13$
Ly = LyS	Lyons-Sims	$2^8 \cdot 3^7 \cdot 5^6 \cdot 7 \cdot 11 \cdot 31 \cdot 37 \cdot 67$
He = HHM	Held-Higman-McKay	$2^{10} \cdot 3^3 \cdot 5^2 \cdot 7^3 \cdot 17$
Ru	Rudvalis	$2^{14} \cdot 3^3 \cdot 5^3 \cdot 7 \cdot 13 \cdot 29$
O'N = O'NS	O'Nan-Sims	$2^9 \cdot 3^4 \cdot 5 \cdot 7^3 \cdot 11 \cdot 19 \cdot 31$
Co_3 = ·3	Conway	$2^{10} \cdot 3^7 \cdot 5^3 \cdot 7 \cdot 11 \cdot 23$
Co_2 = ·2		$2^{18} \cdot 3^6 \cdot 5^3 \cdot 7 \cdot 11 \cdot 23$
Co_1 = ·1		$2^{21} \cdot 3^9 \cdot 5^4 \cdot 7^2 \cdot 11 \cdot 13 \cdot 23$
M(22) = F_{22}	Fischer	$2^{17} \cdot 3^9 \cdot 5^2 \cdot 7 \cdot 11 \cdot 13$
M(23) = F_{23}		$2^{18} \cdot 3^{13} \cdot 5^2 \cdot 7 \cdot 11 \cdot 13 \cdot 17 \cdot 23$

Table 2 (Continued)

Notation	Name	Order
$M(24)' = F_{24}$		$2^{21} \cdot 3^{16} \cdot 5^2 \cdot 7^3 \cdot 11 \cdot 13 \cdot 17 \cdot 23 \cdot 29$
$F_3 = E$	Thompson	$2^{15} \cdot 3^{10} \cdot 5^3 \cdot 7^2 \cdot 13 \cdot 19 \cdot 31$
$F_5 = D$	Harada	$2^{14} \cdot 3^6 \cdot 5^6 \cdot 7 \cdot 11 \cdot 19$
$F_2 = B$	Baby Monster	$2^{41} \cdot 3^{13} \cdot 5^6 \cdot 7^2 \cdot 11 \cdot 13 \cdot 17 \cdot 19 \cdot 23 \cdot 31 \cdot 47$
$F_1 = M$	Monster	$2^{46} \cdot 3^{20} \cdot 5^9 \cdot 7^6 \cdot 11^2 \cdot 13^3 \cdot 17 \cdot 19 \cdot 23 \cdot 29 \cdot 31 \cdot 41 \cdot 47 \cdot 59 \cdot 71$

group on Ω is the group of all permutations on Ω under composition. A permutation group G on Ω is said to be <u>n-transitive</u> if for any pair of ordered n-tuples $(a_1, \ldots, a_n)$ and $(b_1, \ldots, b_n)$ of distinct elements of Ω there exists some $g \in G$ mapping the first n-tuple to the second. That is $a_i g = b_i$ for each $1 \leq i \leq n$, where ag denotes the image of $a \in \Omega$ under g. G is <u>multiply transitive</u> if G is n-transitive for $n \geq 2$. The Mathieu group M_n is 3, 4, or 5 transitive on a set of order n. Indeed the only known finite groups which are as much as 4-transitive are the Mathieu groups, the alternating groups, and the symmetric groups. Presumably there are no others. This is one of the questions that the classification of the simple groups will settle. Presumably, it could be settled in no easier fashion.

As far as I know, the question of the existence and uniqueness of the Mathieu groups was not settled until the 1930's in a paper of Witt [40]. Witt accomplished this by demonstrating the existence and uniqueness of the Steiner system $S(5, 8, 24)$, whose automorphism group is the Mathieu group M_{24}. A Steiner system $S(t, k, n)$ is a set Ω of n points together with a collection B of subsets of Ω, each of size k, such that any t points of Ω are contained in a unique member of B. The other Mathieu groups are also associated to Steiner systems. These Steiner systems and their associated Mathieu groups give rise to many interesting combinatorial objects, such as, perfect codes.

There was a gap of over a century before the next sporadic group J_1 was discovered by Z. Janko in 1965. I will skip on however to the next sporadic group J_2 = HJ discovered after J_1, since one of the methods by which this group was discovered is closely related to multiply transitive permutation groups.

Let G be a transitive permutation group on Ω. We may also regard G as a permutation group on the set product $\Omega \times \Omega$ under $(a,b)g = (ag, bg)$. Let $\sim$ be the equivalence relation on $\Omega \times \Omega$ defined by $x \sim y$ if $xg = y$ for some $g \in G$. The equivalence classes of this relation are called the <u>orbits</u> of G on $\Omega \times \Omega$. The <u>permutation rank</u> of G is the number of orbits of G on $\Omega \times \Omega$. Evidentially one orbit of G is

$$\{(a, a) : a \in \Omega\}$$

so if Ω has at least two points then G has rank at least 2. A moments thought shows G has rank exactly 2 precisely when G is multiply transitive. The next case to consider is the case where G is a rank 3 group. In that case, let Δ and Γ be the two nontrivial orbits of G on $\Omega \times \Omega$. For $\alpha \in \Omega$ define $\Delta(\alpha)$ to consist of those β in Ω with $(\alpha, \beta) \in \Delta$.

As we are interested in simple groups, the Feit-Thompson theorem tells us we may take G to be of even order. Under this restriction it is true that for $\alpha, \beta \in \Omega$, $\alpha \in \Delta(\beta)$ if and only if $\beta \in \Delta(\alpha)$. Hence we may define an undirected graph on Ω by joining α to the points in $\Delta(\alpha)$. Evidently, G acts as a group of automorphisms of this graph. Moreover the graph is <u>strongly regular</u>. That is the number of points $|\Delta(\alpha) \cap \Delta(\beta)|$ adjacent to two points α and β depends only on whether $\alpha = \beta$, $\beta \in \Delta(\alpha)$, or $\alpha \neq \beta \notin \Delta(\alpha)$. One can associate a reasonably nice geometric structure to a strongly regular graph which can be used to investigate the corresponding rank 3 group.

The Hall-Janko group HJ was discovered by Janko; M. Hall then constructed the group as a rank 3 permutation group. The groups HS, Sz, Mc, and Ru were also discovered as rank 3 groups.

Bernd Fischer discovered the four sporadic groups denoted by M(22), M(23), M(24)', and F_2 using his theory of finite groups. This will be discussed in some detail in Chapter III.

John Conway discovered the three sporadic groups denoted by Co_1, Co_2, and Co_3 as subquotients of the automorphism group of the Leech lattice. The Leech lattice is a certain discrete Z-submodule of 24-dimensional Euclidean space related to the Steiner system S(5, 8, 24).

The remainder of the sporadic groups were discovered by studying centralizers of involutions. As we shall see in the next chapter, this approach is an important part of the local theory. The other approaches have little if anything to do with the local theory.

Let G be a group. An _involution_ of G is an element of G of order 2. For $x \in G$ the _centralizer_ in G of x, written $C_G(x)$ or C(x), is the subgroup of G consisting of all elements commuting with x.

Let H be a group and P the property that G possesses an involution x with $C(x) \cong H$. Most sporadic groups have been discovered investigating simple groups satisfying P for various centralizers H. To be specific let H = C(x) with $H/\langle x \rangle \cong F_2$.

F_2 is the sporadic group commonly called the <u>Baby Monster</u>. Fischer, Thompson, Conway, and Harada, and independently Griess, investigated this question. Griess showed that if G is a simple group satisfying P, then G has the following order:

$$2^{46} \cdot 3^{20} \cdot 5^9 \cdot 7^6 \cdot 11^2 \cdot 13^3 \cdot 17 \cdot 19 \cdot 23 \cdot 29 \cdot 31 \cdot 41 \cdot 47 \cdot 59 \cdot 71$$

or about 10^{54}. A simple group satisfying P is called the <u>Monster</u> and denoted by F_1. As mentioned earlier the existence and uniqueness questions for this group are unsettled. The Monster is the largest sporadic group yet discovered.

When the Monster was discovered it was observed that, if the group existed, it must contain two new sporadic groups (the groups denoted by F_3 and F_5 in Table 2) whose existence had not been suspected up to that time. That is, these groups were discovered as subgroups of the Monster. Since that time the groups F_3 and F_5 have been shown to exist. This is analogous to the situation in the physical sciences where a theory is constructed which predicts certain physical phenomena that are later verified experimentally. Such verification is usually interpreted as evidence that the theory is correct. In this case, I take the existence of F_3 and F_5 to be very good evidence that the Monster exists.

I have gone to some length in discussing the discovery of the various sporadic groups for a number of reasons. For one thing, it seems to be the area of finite group theory which most captures the fancy of nonspecialists. There are reasons more relevant to the classification however. First, as I have tried to indicate, some of the groups were discovered through the local theory while others were discovered by methods which have little if anything to do with that theory. Still it is the case that all the groups appear more or less similar from the point of view of the local theory and indeed appear as small pathologies to the theory. On the other hand, while there are marked similarities in the local structure of the groups, no elegant theory is available to describe, at the same time, all or even most of the groups. Hence the groups which were not discovered through the local theory give some verification that the theory is on the right track. Moreover we have some evidence that the sporadic groups are best viewed as small pathologies.

I know of no reference which discusses any sizable number of the sporadic groups. Perhaps Conway's article in [28] comes closest. There you will find a discussion of the Mathieu groups and the Conway groups.

In summary the known nonabelian simple groups consist of the groups of Lie type, the alternating groups, and the sporadic groups. The groups of Lie type are the generic simple groups. The sporadic groups are best viewed as small pathologies. Denote by K the set of all known simple groups. The aim of the classification program is to prove each simple group is isomorphic to a member of K. Possibly in the process one or more new sporadic groups will be discovered which must then be added to K. As long as these groups resemble the known groups from the point of view of the local theory and as long as there are not too many such groups, then we are all right. The existing techniques can be perturbed slightly to allow such extra groups, and to work with the enlarged K. My belief is that there are at most a few groups yet to be discovered. If I were to bet, I would say no more.

Recall also that the uniqueness question for the Ree groups and existence and uniqueness questions for two of the sporadic groups is are unsettled. What we will probably do is prove a theorem which says that if G is a finite simple group, then either G is isomorphic to a member of K or G satisfies the defining properties of a group of Ree type or one of the sporadic groups J_4 or $F_1 \cdot _4$ or F_1.

II. The Local Theory

This chapter gives a quick introduction to the local theory of finite groups and indicates, at the most basic level, how this theory is used in the classification of the simple groups. Later chapters will describe the classification program itself.

The local theory of finite groups studies the collection of <u>local subgroups</u> of a finite group. Here given a prime p, a <u>p-local subgroup</u> of a group G is the normalizer of a nontrivial p-subgroup of G. The centralizers of elements of prime order are particularly important. The prime 2 plays a special role.

The special behavior of the prime 2 pervades every aspect of finite group theory. It will appear again and again in each subproblem discussed. In particular, local pathologies seem to occur only for small primes, with the frequency and severity of the pathologies increasing as the size of the prime decreases.

To illustrate concepts to be discussed in this chapter, let us consider an example. Let $GL_n(q)$ be the n-dimensional general linear group over the field of order q. That is, $GL_n(q)$ is the group of all n by n nonsingular matrices with entries over $GF(q)$. I believe that $GL_n(q)$ is the best group to consider to build up some intuition about the finite simple groups. This is true even though $GL_n(q)$ is not usually simple. But this is no particular problem as the group is very close to being simple. To be more precise, the group $SL_n(q)$ of all n by n matrices of determinant 1 over $GF(q)$ is the <u>special linear group</u> and $L_n(q)$ is $SL_n(q)$ modulo the scalar matrices of determinant 1. $L_n(q)$ is usually simple. Indeed from Table 1, $L_n(q)$ is isomorphic to

$A_{n-1}(q)$, a group of Lie type. It is advantageous to consider $GL_n(q)$ rather than $L_n(q)$ because it is easier to make calculations in the former group. The reader should think of $GL_n(q)$ as a group of Lie type over $GF(q)$. $GL_n(q)$ will be our generic simple group.

We will be interested in two types of elements in $G = GL_n(q)$: the semisimple and unipotent elements. A matrix x in G is <u>semisimple</u> if x is diagonalizable over some extension of $GF(q)$. x is <u>unipotent</u> if $x - I$ is nilpotent, where I is the identity matrix. Let q be a power of the prime p. Then unipotent elements are <u>p-elements</u> and semisimple elements are <u>p'-elements</u>. That is if x is unipotent then the order of x is a power of p, while if x is semisimple then the order of x is relatively prime to p. Each element of G may be written uniquely as a product of commuting semisimple and unipotent elements.

Let us consider the centralizer of typical semisimple and unipotent elements. Figures 1 and 2 show the centralizers of a semisimple element s and a unipotent element u, respectively.

s has been chosen to have two distinct eigenvalues a and b with eigenspaces of dimension r and $n - r$, respectively. The centralizer of s is then the direct product of the general linear groups on these eigenspaces. Hence the centralizer of a semisimple element is essentially the direct product of "almost simple" subgroups.

The centralizer of u is quite different. The structure of the centralizer of u is dominated by the group U. U is the <u>unipotent radical</u> of $C(u)$. That is U is the largest normal unipotent subgroup of $C(u)$. From the point of view of abstract group theory, U is the largest normal p-subgroup of $C(u)$.

Figure 1

The Centralizer of a Semisimple Element

$$G = GL_n(q)$$

$$s = \begin{pmatrix} aI_r & 0 \\ 0 & bI_{n-r} \end{pmatrix}$$

$$C(s) = \left\{ \begin{pmatrix} X & 0 \\ 0 & Y \end{pmatrix} : X \in GL_r(q), Y \in GL_{n-r}(q) \right\} \cong GL_r(q) \times GL_{n-r}(q)$$

Figure 2

The Centralizer of a Unipotent Element

$$G = GL_n(q)$$

$$u = \begin{pmatrix} I_r & 0 & 0 \\ 0 & I_{n-2r} & 0 \\ I_r & 0 & I_r \end{pmatrix}$$

$$Cu = \left\{ \begin{pmatrix} X & 0 & 0 \\ P & Y & 0 \\ Q & R & X \end{pmatrix} : \begin{array}{l} X \in GL_r(q),\ Y \in GL_{n-2r}(q) \\ P,\ Q,\ R \text{ arbitrary} \end{array} \right\}$$

$$U = \left\{ \begin{pmatrix} I_r & 0 & 0 \\ P & I_{n-2r} & 0 \\ Q & R & I_r \end{pmatrix} : P,\ Q,\ R \text{ arbitrary} \right\}$$

$$= O_p(C(u)),\quad q = p^e$$

= Unipotent Radical of $C(u)$

$$C(u)/R \cong GL_r(q) \times GL_{n-2r}(q)$$

In general if π is a set of primes and H a group, $O_\pi(H)$ denotes the largest normal subgroup of H whose order is divisible only by primes in π. $O_p(H)$ is the largest normal p-subgroup of H.

We see then the centralizers of semisimple elements are essentially the direct product of "almost simple" subgroups while the centralizers of unipotent elements are dominated by large normal p-subgroups. We wish to make these concepts precise and to phrase them in the language of abstract group theory rather than the theory of linear or algebraic groups.

Define a group L to be **quasisimple** if $L = [L,L]$ is perfect and $L/Z(L)$ is simple. L is **subnormal** in a group G if there exists a series of subgroups:

$$L = G_o \trianglelefteq G_1 \trianglelefteq \ldots G_n = G$$

A component of G is a subnormal quasisimple subgroup of G. It develops that distinct components of G commute so that the group $E(G)$ generated by all components of G is the direct product of these components modulo its center. Of course G permutes the components under conjugation.

In Figure 1 the components of the centralizer of the semisimple element s are essentially the GL(q) factors. To be precise, the components are the SL(q) factors, as $SL_n(q)$ is quasisimple unless $n = 2$ and $q \leq 3$. The centralizer of the unipotent element in Figure 2 has no components.

We see then that the centralizer of a generic semisimple element possesses components and the structure of the centralizer is dominated by these components. On the other hand, the centralizer of a unipotent p-element has no components, rather its structure is dominated by a large normal p-subgroup.

Finite group theorists often draw diagrams similar to those in Figure 3 to represent groups. Here $1 \trianglelefteq G_1 \trianglelefteq G_2 \trianglelefteq G$ is a series of normal subgroups with the boxes indicating the factors induced by the series. We tend to think of the factor G_1 as at the bottom of the group and the factor G/G_2 as at the top. From this point of view components correspond to composition factors near the bottom of the group. These factors seem to be easier to work with and have more control of the group structure than factors nearer the top.

We will say that the centralizer of an element x is in <u>standard form</u> if $C_G(x)$ has one component L and $C_G(L)$ is small in some sense. Typically, for example, x will be of prime order p and $C_G(L)$ will have cyclic Sylow p-groups. Such a centralizer is essentially embedded in the automorphism group of the simple group $L/Z(L)$ and hence is reasonably well determined.

If G is the general linear group $GL_n(q)$, choose s to be a semisimple element with an eigenspace of dimension $n - 1$. Then as in Figure 1

$$C_G(s) \cong GL_1(q) \times GL_{n-1}(q)$$

with $GL_1(q)$ cyclic of order $q - 1$, and $C(s)$ is in standard form with component $SL_{n-1}(q)$.

The fundamental principle underlying the classification theory of finite simple groups is that a simple group can be determined from the centralizer of an element in standard form. To establish that a given simple group is determined by a given centralizer in standard form is usually not too hard. The more difficult problem is to show that an arbitrary finite simple group possesses such a centralizer.

Figure 3

A Group Diagram

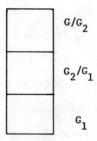

$1 \unlhd G_1 \unlhd G_2 \unlhd G$

For the most part I must ask you to take the fundamental principle on faith. I will make just one remark in attempt to convince you of its validity. While it seems possible to classify the finite simple groups, it seems out of the question to classify finite groups in general. There is an incredible variety of p-groups for example. Thus we must make effective use of the hypothesis of simplicity. The existence of a centralizer in standard form has the effect of forcing a group to be essentially simple. For example:

Lemma 2.1. Let G be a finite group with $O(G) = 1$ and x an involution in G such that $C_G(x)$ has one component L and $C_G(L)$ has cyclic Sylow 2-subgroups. Then either

   (1) $G = C_G(x)$

   (2) $H \leq G \leq \text{Aut}(H)$ for some simple group H,

   (3) $L \times L \leq G \leq \text{Aut}(L \times L)$ and L is simple.

Here $O(G)$ is the largest normal subgroup of G of odd order and $\text{Aut}(G)$ is the automorphism group of G. An involution is an element of order 2. Similar results hold for elements of order $p > 2$.

The point here is that if we have a centralizer in standard form, then we know right away that our group is essentially simple and it only remains to determine which simple group is before us. We have eliminated from consideration almost all finite groups and are left with a relatively small number of groups to consider.

Our generic finite simple group $GL_n(q)$ has an associated characteristic: the characteristic p of the field of definition. Our next job is to associate a characteristic with an abstract finite simple group. We have seen that the semisimple elements in $GL_n(q)$ are p'-elements while the unipotent elements are p-elements. The centralizers of semisimple elements have components while if u is a unipotent element in $L_n(q) = G$ then

$$C(u) \cap C(O_p(C(u))) \leq O_p(C(u)).$$

Thus we define an abstract group G to be of <u>characteristic p-type</u> if

$$C_H(O_p(H)) \leq O_p(H)$$

for each p-local subgroup H of G. This concept will be useful for the prime p only when the collection of p-subgroups of G is sufficiently rich. For example, if p is not a prime divisor of G then G possesses no p-locals and hence is vacuously of characteristic p-type. Also a group G may have no characteristic or more than one characteristic. The group $^2A_3(2) \cong C_2(3)$ is of characteristic 2 type and characteristic 3 type in a nontrivial fashion. All groups of Lie type over fields of characteristic p are of characteristic p-type.

If G is of characteristic p-type it is usually difficult to use centralizers of elements of order p to determine G. This is because the structure of such a centralizer is dominated by its large normal p-subgroup and p-groups are notoriously difficult to deal with. Following the fundamental principle, we search instead for centralizers in standard form.

As mentioned earlier, the prime 2 plays a special role in the theory of finite simple groups. Hence, one first attempts to show the centralizer of some involution is in standard form. For example, this will usually be the case if G is of Lie type of odd characteristic or an alternating group. It will also be the case for many sporadic groups.

Whenever the sporadic groups are encountered, we should look closely to see what exceptional behavior is at work. What we find is that if G is sporadic and L is a component of the centralizer of some involution in standard form, then $L/Z(L)$ is a group of Lie type over $GF(2)$ or $GF(4)$ of small Lie rank or a small sporadic or alternating group. Now, if G were of Lie type and characteristic $p > 2$, then the components of centralizers of involutions would be of Lie type in characteristic p, while if G were of Lie type and even characteristic, then involutions would be unipotent and their centralizers possess no components. Thus we see these sporadic groups emerging from the local theory as groups with involutions whose centralizers are in some unusual standard form.

In the case where a simple group G has no involution whose centralizer is in standard form, it seems possible to show that, modulo a few known exceptions, that G is of characteristic 2-type. In this case, we look for elements of odd prime order whose centralizers are in standard form. However, the fact that G is of characteristic 2-type will be of much help in this analysis.

The known simple groups of characteristic 2-type include the groups of Lie type of even characteristic, a few groups of Lie type over $GF(3)$, and some of the sporadic simple groups. In the sporadic case, elements of prime order whose centralizers are in standard form will tend to be of order p with components of Lie type over $GF(p)$ which are also of characteristic 2-type, or small sporadic components.

Moreover, the 2-local structure is reminiscent of a group of Lie type over GF(2).

If G is not of characteristic 2-type, it can be shown that with known exceptions, G possesses an involution x such that $C(x)/O(C(x))$ has a component. Such a group is said to be of <u>component type</u>. The problem is then to show that a group of component type possesses an involution whose centralizer is in standard form and that a group of characteristic 2-type, modulo known exceptions, possesses an element of odd prime order whose centralizer is in standard form.

In summary, from the point of view of the local theory, there is a basic subdivision of the finite simple groups into the groups of characteristic 2-type and the groups of component type. The analysis in these two cases is quite different. The basic flow diagram is indicated in Figure 4. A discussion of the two subdivisions appears in Chapters IV, V, and VI.

At this time, the groups of component type seem to be better understood then those of characteristic 2-type. We have the feeling that the groups of component type are all but classified. On the other hand, much work is left to be done to classify the groups of characteristic 2-type, although here too we feel we have a good idea of just what to do.

Figure 4

Flow Diagram for the Classification

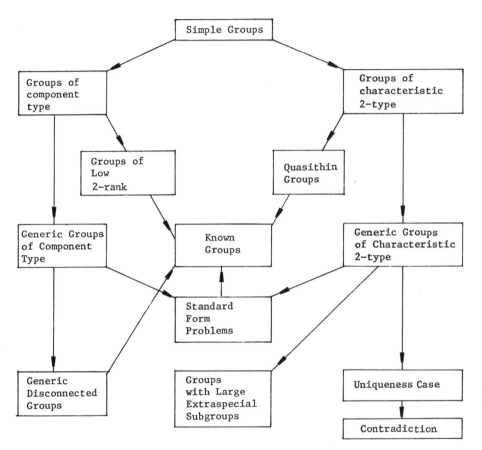

This illustration provides a flow diagram of the classification.
The various steps will be discussed in later chapters. The
reader may wish to refer to this diagram as these steps are
encountered.

III. Fischer's Theory of Finite Groups

Bernd Fischer originated a theory of finite groups about ten years ago. Since that time the theory has proved to be useful in the study of finite simple groups, although it is practiced by relatively few group theorists.

Let G be a finite group and D a collection of subgroups or elements of G permuted by G under conjugation. That is D is a union of conjugacy classes of subgroups or elements. It is also convenient to take G to be generated by D. It is Fischer's idea to consider the situation where for each pair of members A and B from D, the isomorphism class of the subgroup $<A,B>$ generated by A and B is in some small set of groups.

For example, let G be a group of Lie type over GF(q). There is a conjugacy class D of subgroups of G known as long root subgroups which are intimately connected to the (B,N)-pair structure of G. Usually D satisfies the following conditions:

For each pair of distinct members A and B in D, A is an elementary abelian p-group of order $q = p^e$ and $X = <A,B>$ satisfies one of the following:

(3.1)   X is elementary abelian of order q^2.

(3.2)   X is isomorphic to $SL_2(q)$.

(3.3)   X is of order q^3 with $[A,B] \in D$.

An elementary abelian p-group is an abelian p-group in which each nontrivial element has order p.

More generally, if G is a finite group generated by a G-invariant set D of subgroups satisfying the properties above, we shall say that D is a <u>root generating</u> set for G.

Using the theory developed by Fischer and his various students and disciples, plus some local theory, the groups with a root generating set have been determined, except for a small gap when $q = 3^e > 3$. See, for example, [34], [3], [10], [8], [37], [1], and [14]. Most of these groups are of Lie type over $GF(q)$.

The classification of groups with a root generating set for even q has proved to be an important tool in the classification program, particularly in the classification of groups of characteristic 2-type. Presumably, a classification of these groups would also be quite useful in investigating the subgroups of groups of Lie type. For example, when G is of Lie type and H a subgroup of G with $H \cap D$ nonempty, then $< H \cap D>$ would be a known normal subgroup of H.

The original problem which Fischer considered was much more specialized. Namely, let us consider root generating sets with $q = 2$ and assume in addition that case 3.3 does not occur. This will hold for example in the symplectic and unitary groups over $GF(2)$. Then we may regard D as a G-invariant set of involutions, generating G, and with the property that for each a and b in D, the order of ab is at most 3. Fischer calls such a set of involutions a set of <u>3-transpositions</u>. Observe that the ordinary transpositions in the symmetric group form a set of 3-transpositions. It was Fischer's original intention to characterize the symmetric groups from this point of view. In the process, he was lead not only to his theory of finite groups, but also to three sporadic groups, denoted by $M(22)$, $M(23)$, and $M(24)'$ in Table 2. These groups are generated by 3-transpositions. A little later he was lead to discover a fourth sporadic group, the Baby Monster, denoted by F_2 in Table 2. The Baby Monster is generated by $\{3,4\}$-transpositions. That is for each a and b in D, the order of ab is at most 4.

Fischer's approach is to associate some geometric structure to the group. Namely, define $G = \mathcal{D}(G,D)$ to be the graph with vertex set D and A adjacent to B if $A \neq B$ and $[A,B] = 1$, (ie. A and B commute). Then G is an undirected graph with no loops and G is represented as an automorphism group of the graph by conjugation. We wish to make use of this fact.

In the following discussion, I adopt a point of view somewhat different from Fischer's original approach, taken rather from [5].

For the moment, let us forget about our group and assume only that G is an undirected graph with no loops. Let D be the vertex set of G and for $A \in D$ let $G(A)$ be the set of vertices adjacent to A in G. Set $A^\perp = \{A\} \cup G(A)$. Let G^c be the <u>complementary graph</u> of G. That is, G^c is the graph with vertex set D and $A \in D$ adjacent to the vertices in $G^c(A) = D - A^\perp$.

A <u>homomorphism</u> of a graph G_1 onto a graph G_2 is a map α of the vertex set D_1 of G_1 onto the vertex set D_2 of G_2 which preserves adjacency in G_1 <u>and</u> G_1^c. That is, if $A \in D$, $B \in G_1(A)$, and $C \in G_1^c(A)$ then $B\alpha \in G_2(A\alpha) \cup \{A\alpha\}$ and $C\alpha \in G_2^c(A\alpha) \cup \{A\alpha\}$. Observe, that α corresponds to a partition

$$\Gamma = \{A\alpha^{-1} : A \in D_2\}$$

of D_1 with the property that if $X, Y \in \Gamma$, $X \neq Y$, $A \in X$, $B \in Y$, and $A \in G_1(B)$ then $Y \subseteq G_1(C)$ for each $C \in X$. Such partitions will be called <u>contractions</u> and the members of the partition will be called <u>clumps</u>. Hence α contracts the clumps of G_1 to points of G_2. Exclude the partition $\Gamma = \{D\}$ from being a contraction.

Conversely, if Γ is a contraction of a graph G let G/Γ be the graph with vertex set Γ and X adjacent to Y if there is $A \in X$ and $B \in Y$ with A adjacent to B in G. Then the inclusion map is a homomorphism of G onto G/Γ.

Partially order contractions by $\Gamma_1 < \Gamma_2$ if each clump of Γ_1 is contained in some clump of Γ_2. If $\Gamma_1 < \Gamma_2$ then Γ_2 may be regarded as a contraction of G/Γ_1 and hence G/Γ_2 is a homomorphic image of G/Γ_1. Thus the maximal contractions are of most interest. Define G to be <u>simple</u> if G has no nontrivial homomorphic images. Then Γ is a maximal contraction precisely when G/Γ is simple.

The following results are in [5]:

Theorem 3.4. Let G and G^c be connected. Then G has a unique maximal contraction and hence a unique simple homomorphic image.

Theorem 3.5. Let G and G^c be connected and G simple. Let $A \in D$. Then $G(A)^c$ and $(G^c(A))^c$ are connected.

Let us return now to the case where $G = \mathcal{D}(G,D)$. Then G^c is disconnected if and only if D admits a nontrivial partition $D = D_1 + D_2$ with $[D_1, D_2] = 1$. In this case, $G = <D>$ is the central product of the groups $<D_1>$ and $<D_2>$. Thus the hypothesis that G^c be connected is not such a strong one. The question of when G is connected is much deeper. Many important problems in finite group theory revolve around this question. See the discussion in Chapter IV.

Assume then that G and G^c are connected and that G is transitive on D. For $A \in D$ let $\alpha(A)$ be the unique maximal clump containing A. Then $\Gamma = \alpha(A)^G$ is a partition of D with G transitive on Γ. If we choose, we may replace D by

$$\{<\alpha(A)> : A \in D\}$$

to get G simple. Or we may be able to prove G is simple.

It is time to begin putting restrictions on the groups generated by pairs from D. For example, define D to be <u>locally conjugate</u> in G if for A and B in D either $[A,B] = 1$ or A is conjugate to B in $<A,B>$. Notice that if D is a set of 3-transpositions, then D is locally conjugate. Evidentially:

Lemma 3.6. Let D be locally conjugate in G. Then G^c is connected if and only if G is transitive on D.

As a corollary to this remark and Theorem 3.5, we have:

Theorem 3.7. Let D be locally conjugate, G and G^c connected, G simple, $A \in D$, and $B \in G(A)$. Then

(1) $<G(A)>$ is transitive on $G(A)$
(2) $G = <G(A), G(B)>$

We are already generating useful group theoretic information.

A <u>triangle</u> in D is a triple (A, B, C) with $C \in G(A)$ and $A, C \in G^c(B)$. G has the <u>triangle property</u> if for each triangle (A, B, C), C is conjugate to A in $<<A, B, C> \cap B^\perp>$. It turns out that a set D of 3-transpositions has the triangle property. For (a, b, c) is a triangle precisely when $X = <a, b, c>$ is a

Coxeter group with reflections a, b, c such that $(ab)^3 = (bc)^3 = (ac)^2 = 1$. Thus X is isomorphic to the symmetric group on 4 letters and we check easily that the triangle property is satisfied.

It is now immediate from Theorem 3.5 and 3.7 that:

Theorem 3.8. Let G and G^c be connected, G simple, D locally conjugate with the triangle property, and $A \in D$. Then $< A^\perp >$ is transitive on $G(A)$ and $G^c(A)$. In particular, G acts as a rank 3 permutation group on G and G is a strongly regular graph.

I will show in a moment, that if G is an interesting group generated by 3-transpositions, then G and G^c are connected and G is simple. Hence the graph of an interesting 3-transposition group is strongly regular. As remarked in Chapter I, such groups have much geometric structure. Fischer was able to utilize this structure to go on and classify 3-transposition groups. His theorem has since been extended, using similar methods, to a classification of groups with a root generating set.

Suppose now that G is transitive on D. A subset Δ of D is a __set of imprimitivity__ for G if the set of G-conjugates of partitions D. For example, if G and G^c are connected and Δ a maximal clump then Δ is a set of imprimitivity. If G is disconnected and Δ a connected component of G then again Δ is a set of imprimitivity.

The following lemma is very easy:

Lemma 3.9. Let D be locally conjugate and Δ a set of imprimitivity. Then Δ^G is a contraction of G.

Theorem 3.10. Let D be a conjugacy class of 3-transpositions in G and $[G,G]$ simple. Then G and G^c are connected and G is simple.

The proof is not hard if we assume $a \in D$ commutes with some other member of D. We may assume this by Glauberman's Z^*-theorem [16]. First, as $[G,G]$ is simple and $G = <D> = <a^G>$, $G = [G,G] <a>$. Next by Lemma 3.6, G^c is connected, so by a remark above, we may take a to be properly contained in a set of imprimitivity $\Delta(a) = \Delta$. Let $a \neq c \in \Delta(a)$ and $b \in D$. We show ac fixes $\Delta(b)$, and hence fixes Δ^G pointwise. Hence as $[G,G]$ is simple, $[G,G]$ fixes Δ^G pointwise and then $G = [G,G] <a>$ fixes Δ, a contradiction.

So assume ac does not fix $\Delta(b)$. Then $[ac, \Delta(b)] \neq 1$ so by Lemma 3.9, ab and ac have order 3. Thus $a^b = b^a$ and $b^c = c^b$. Then $\Delta(b)^a = \Delta(b^a) = \Delta(a^b) = \Delta(a)^b = \Delta(c)^b = \Delta(c^b) = \Delta(b^c) = \Delta(b)^c$ and ac fixes $\Delta(b)$.

IV. Signalizer Functors and Disconnected Groups

In the remainder of this manuscript, I will attempt to describe some of the high points in the program to classify the finite simple groups. Let G be a finite simple group and x an element of G of prime order p. We wish to show $C_G(x) = H$ to be in standard form. The first obstruction we encounter, and indeed possibly the largest obstruction, is the group $O_{p'}(H)$. Recall $O_{p'}(H)$ is the largest normal subgroup of H of order relatively prime to p. We wish to show $O_{p'}(H)$ to be small in some sense. For example, in the known simple groups $O_{p'}(H)$ seems to be cyclic. However, for most purposes the following condition seems to be the crucial one:

B_p-Conjecture. Let G be a group with $O_{p'}(G) = 1$. Then, for each element x of order p, $E(C(x)/O_{p'}(C(x))) = E(C(x))O_{p'}(C(x))/O_{p'}(C(x))$.

That is, each component of $C(x)/O_{p'}(C(x))$ is covered by a component of $C(x)$. The analogous property for algebraic groups might be that the centralizers of semisimple elements in a reductive algebraic group are reductive. From this point of view, $O_{p'}(H)$ is analogous to the radical of H.

As usual, the case $p = 2$ is of most interest and we refer to the B_2-conjecture as the <u>B-conjecture</u>. The B-conjecture appears to be all but established. We will see just how close in the next chapter. For odd primes it seems best to finesse the conjecture rather than to establish it directly.

The first tool for analyzing $O_{p'}(H)$ is the concept of the <u>signalizer</u> <u>functor</u>. This concept is due to Daniel Gorenstein and is suggested by analysis in the Odd Order Paper and the N-group Paper.

Recall that an elementary abelian p-group is an abelian p-group in which each nontrivial element is of order p. Such a group may be regarded as a vector space over $GF(p)$; the _rank_ of the group is the dimension of this vector space.

Assume A is an elementary abelian p-subgroup of G of rank at least 3. An _A-signalizer functor_ is a map θ which associates to each $a \in A^\# = A - \{1\}$ an A-invariant subgroup $\theta(a)$ of $C(a)$ of order relatively prime to p, such that for each $a, b \in A^\#$ the following condition, known as _balance_, holds:

$$\theta(a) \cap C(b) = \theta(b) \cap C(a)$$

Here is the basic example of a signalizer functor. Indeed we will see soon that it is the only example. Let G be a group, A an elementary abelian p-subgroup of G of rank at least 3, and X an A-invariant p'-subgroup of G. For example, X might be $O_{p'}(G)$. Define

$$\theta(a) = C_X(a)$$

Then evidently, θ is an A-signalizer functor and $\langle \theta(a) : a \in A^\# \rangle$ is a subgroup of X, and hence a p'-group. Actually more is true. The following elementary lemma appears as 6.2.4 in [20]:

Lemma 4.1. Let B be a noncyclic elementary abelian p-group acting on a p'-group Y. Then

$$Y = \langle C_Y(D) : D \leq B, |B:D| = p \rangle$$

Hence $X = \langle \theta(a) : a \in B^\# \rangle$ for each noncyclic subgroup B of A.

Now, if $X = O_{p'}(G)$ we have a description of X in terms of subgroups $\theta(a)$ of $C(a)$, $a \in A^{\#}$. That is, we have a p-local description of $O_{p'}(G)$. Suppose further, $X = 1$. Then we expect $O_{p'}(C(a))$ to be small for each $a \in A^{\#}$. We know $\theta(a) \leq X = 1$, and we hope to be able to use this information to prove $O_{p'}(C(a))$ small.

This is the plan; we proceed with its implementation. Given an A-signalizer functor θ and a subgroup B of A define

$$\theta(B) = <\theta(b): b \in B^{\#}>.$$

If we have chosen θ wisely, then $\theta(A)$ should be essentially $O_{p'}(G)$. As a first step, we show $\theta(A)$ to be a p'-group. This is accomplished in the following theorem:

Signalizer Functor Theorem. Let θ be an A-signalizer functor and assume for each $a \in A^{\#}$ that $\theta(a)$ is solvable, or more generally, that all composition factors of $\theta(a)$ are known simple groups. Then $\theta(A)$ is a p'-group, $\theta(A) \cap C(a) = \theta(a)$ for each $a \in A^{\#}$, and $\theta(A) = \theta(B)$ for each noncyclic subgroup B of A.

I emphasize that A must be of rank at least 3 or else there are counter examples to the signalizer functor theorem. A number of mathematicians have contributed to the proof. Gorenstein originated the signalizer functor concept and proved the first weak forms of the signalizer functor theorem [21, 22]. θ is said to be __solvable__ if $\theta(a)$ is solvable for each $a \in A^{\#}$. If $p = 2$, then $\theta(a)$ is of odd order by hypothesis and hence solvable by the Feit-Thompson Theorem. Goldschmidt was the first to prove the solvable signalizer functor theorem for $p = 2$. [19] He also proved a solvable signalizer functor theorem for odd primes in case A is of rank at least 4. [18]. Glauberman proved the general solvable signalizer functor theorem [17]. Gorenstein and Lyons did the first work on nonsolvable signalizer functors; McBride extended their work to establish the signalizer functor theorem under the hypothesis that the composition factors of the groups θ(a) are of known type. If we are working in a simple group minimal subject to not being known, then this hypothesis holds by induction.

The next step is to show that θ(A) is normal in G. Define the <u>p-rank</u> $m_p(G)$ of G to be the maximum rank of an elementary abelian p-subgroup of G. Let us assume θ has various extra properties. First assume θ is defined on each element a of G of order p with $m_p(C(a)) > 2$ and that for each elementary abelian p-subgroup A of G of rank at least 3, θ restricted to A is an A-signalizer functor. Assume also that for $g \in G$, $\theta(a^g) = \theta(a)^g$. Here X^g is the conjugate $g^{-1} X g$ of X under g. For example, if $X = O_{p'}(G)$ and $\theta(a) = C_X(a)$ then θ will possess these properties.

To make use of these properties we need some new terminology. Let Ω be the set of noncyclic elementary abelian p-subgroups of G. Recall from Chapter III that $\mathcal{D}(G, \Omega)$ is the graph with vertex set Ω and A adjacent to B if [A,B] = 1. G is said to be <u>connected</u> (for the prime p) if the graph $\mathcal{D}(G,\Omega)$ is connected.

By the signalizer functor theorem, θ(B) = θ(BD) = θ(D) whenever B and D are members of Ω adjacent in $\mathcal{D}(G,\Omega)$. Hence if G is connected, then θ(A) is independent of the choice of A in Ω. Now, if $g \in G$ then $\theta(A) = \theta(A^g) = \theta(A)^g$, so that θ(A) is normal in G.

We are led to the problem of determining all disconnected groups.

The generic group disconnected for the prime p is a group of Lie type of characteristic p and Lie rank 1. In these groups a stronger property is satisfied. Namely, taking Γ to be the set of all nontrivial p-subgroups of G, the graph $\mathcal{D}(G,\Gamma)$ is disconnected. Equivalently, if M is the normalizer in G of a connected component of $\mathcal{D}(G,\Gamma)$, then for each $g \in G - M$, $M \cap M^g$ is a p'-group, while the order of M is divisible by p. A proper subgroup of G with these properties is said to be strongly p-embedded in G. Once again the case $p = 2$ is special and strong 2-embedding is abbreviated to strong embedding.

Groups with a strongly embedded subgroup were classified by Bender in [11], extending earlier work of Suzuki [35]. However, the arguments in this classification depend on very special properties of the prime 2 and do not seem to extend to odd primes. Ultimately, the proof is based upon a beautiful counting argument due to Walter Feit, which depends in turn upon the fact that pairs of involutions generate dihedral groups.

Groups disconnected for the prime 2 are classified in papers by the author [2], Gorenstein and Harada [23], and MacWilliams [27].

The classification of groups with a strongly p-embedded subgroup for odd primes p seems to be one of the most difficult steps left in the program to classify the simple groups. Actually, such a classification appears to be necessary only for groups of characteristic 2-type and only in the case where the strongly p-embedded subgroup is a 2-local. There are no known examples of such groups. This problem will be discussed in Chapter VI.

One needs a classification of groups with a strongly p-embedded subgroup, or groups satisfying some related condition, in many group theoretical problems. Recall that the problem of determining when $\mathcal{D}(G,D)$ is disconnected for certain G-invariant subsets D of G came up in Chapter III. I mention one more example.

Assume $C(x)$ is in standard form for some element x or order p. Methods are available to use this information to determine the isomorphism class of the subgroup

$$M = <C(x^g): x^g \in C(x), g \in G>$$

For example, one might show that M possesses a (B,N)-pair. We expect that M = G, so that G is determined. If this is not the case, then we usually get $\mathcal{D}(G, x^G)$ disconnected or possibly even M strongly p-embedded in G.

V. Groups of Component Type

Assume now that G is simple of component type. There are two points which were not discussed in the last chapter which will be considered here. First, how are suitable signalizer functors constructed? Second, once it is established that G satisfies the B-conjecture, how does one produce centralizers in standard form?

Let A be an elementary abelian 2-subgroup of G of rank at least 3. The construction of A-signalizer functors will depend upon the nature of the components of $C(a)/O(C(a))$, $a \in A^{\#}$, where $O(C(a)) = O_{2'}(C(a))$. For example, the following lemma is relevant:

Lemma 5.1. Either $\theta(a) = O(C(a))$ is an A-signalizer functor or for some $a, b \in A^{\#}$, there is a b-invariant component $\bar{L}$ of $\bar{C}(a) = C(a)/O(C(a))$ such that

$$O(\text{Aut}_{\bar{C}(a)}(\bar{L})) \cap C(\bar{b})) \neq 1$$

Here $\text{Aut}_X(Y) = N_X(Y)/C_X(Y)$ is the group of automorphisms induced on Y in X for any group X and subgroup Y of X.

This leads to the following concept:
A group G is <u>balanced</u> if $O(C(x)) \leq O(G)$ for each involution $x \in G$. I leave it as an exercise for the reader to use the machinery in the last chapter to show that if G is connected for the prime 2 then G is balanced precisely when the functor $\theta(a) = O(C(a))$ is balanced. Together with lemma 5.1 this implies:

Theorem 5.2. Let G be connected for the prime 2. Then either G is balanced or there exists an involution $x \in G$ and a component $\bar{L}$ of $\bar{C}(x) = C(x)/O(C(x))$ such that $\text{Aut}_{\bar{C}(x)}(\bar{L})$ is unbalanced.

This leads in turn to the following important conjecture:

Unbalanced Group Conjecture. Let $L \leq G \leq \text{Aut}(L)$ with L simple and G unbalanced. Then L is one of the following:

(1) A group of Lie type of odd characteristic
(2) An alternating group of odd degree
(3) $L_3(4)$ or He.

If G is balanced with $O(G) = 1$, then $O(C(x)) = 1$ for each involution $x \in G$, so evidently G satisfies the B-conjecture. On the other hand, if $L \leq G \leq \text{Aut}(L)$ with L on the list of the Unbalanced Group Conjecture, then by inspection G satisfies the B-conjecture. Hence, the Unbalanced Group Conjecture implies the B-conjecture.

A remark is in order here. Most results in finite group theory are established by induction. Hence it is often difficult to establish a theorem to which induction cannot effectively be brought to bear. This might occur, for example, if the hypothesis of the theorem does not inherit to enough subgroups and homomorphic images. In such a case, it often becomes necessary to prove a stronger result which does not lend itself to an inductive approach and then to derive the original result as a corollary.

Here we find it necessary to establish the Unbalanced Group Conjecture and derive the B-conjecture as a Corollary. Theorem 5.2 gives us subgroups to which we may apply induction. Namely, if G is a nontrivial counter example to the Unbalanced Group Conjecture, then Theorem 5.2 produces an involution x and a quasisimple group $\overline{L}$ which is known by induction. We hope to use this knowledge to determine G. As we will quickly see, it seems necessary to extend this process yet another step. That is, we will derive the Unbalanced Group Conjecture as a corollary to an even stronger result.

In focusing attention on the Unbalanced groups, we are really focusing on the situations where the map $\theta(a) = O(C(a))$ fails to be a signalizer functor. This map is a first level approximation to the "right" functor. In analyzing unbalanced groups, it is necessary to consider more sophisticated maps which provide higher level approximations. These maps are rather technical and will not be discussed here, except to remark again that the construction depends upon properties of the components of the groups $C(a)/O(C(a))$ for involutions $a \in G$.

Here is the stronger conjecture which implies the Unbalanced Group Conjecture and the B-conjecture:

Component Group Conjecture. Let H be a simple group, $H \leq G \leq \text{Aut}(H)$, x an involution in G and $\bar{L}$ a component of $\bar{C}(x) = C(x)/O(C(x))$ of known isomorphism type. Then H is a known simple group.

This conjecture has several advantages. First, to classify the simple groups it is necessary to establish such a conjecture anyway. Second, in attempting to establish the Unbalanced Group Conjecture, we are led to an analysis of certain proper subgroups which we need to know satisfy the Component Group Conjecture. The Unbalanced Group Conjecture is not strong enough to yield this information inductively; the Component Group Conjecture is sufficiently strong.

Let G be a minimal counter example to the Component Group Conjecture. We first try to show G is balanced. If not, we obtain an involution x and a quasisimple group $\bar{L}$ on the list of the Unbalanced Group Conjecture. Most of the groups on that list are of Lie type of odd characteristic. Let us assume we have reduced to that case. We observe that almost always L possesses an involution z such that $C_L(z)$ has a subnormal subgroup K with $z \in K$ and $K/O(K) \cong SL_2(q)$. We use this fact to attempt to show K is subnormal in $C_G(z)$ and then appeal to the following result of the author [8]:

Theorem 5.3. Let H be simple, $H \leq G \leq \text{Aut}(G)$, x an involution in G, L a subnormal subgroup of $C(x)$ with $x \in L$ and $L/O(L) \cong SL_2(q)$, q odd. Then G is of Lie type and odd characteristic or M_{11}.

Assume now that G is a group of component type satisfying the B-conjecture. We are left with our second question: how to produce a centralizer in standard form? The following reasonably elementary result is crucial; I believe it was first observed by Gorenstein and Walter; see for example [24].

L-balance: Let G be a group in which components of $G/O(G)$ are covered by components of G, let x be an involution in G, and L a component of $C(x)$. Then there exists a component K of G such that either

(1) $L \leq K$, or

(2) $L \leq KK^x$, $K^x \neq K$, and L is a homomorphic image of K.

One then considers the set L of components of centralizers of involutions of G. Define a relation $\longrightarrow$ on L by $L \longrightarrow K$ if there exist commuting involutions x and y in G with L a component of $C(x)$, K a component of $C(y)$ and $L \leq KK^x$.

Let $\mathcal{L}^*$ consist of those L in $\mathcal{L}$ with the property that L is a homomorphic image of K whenever $L \longrightarrow K$. For example, if L is a member of $\mathcal{L}$ of maximal order then L is in $\mathcal{L}^*$.

The following result says that the members of $\mathcal{L}^*$ are usually components of centralizers in standard form. The theorem is due to the author and R. Foote and can be found in [4] and [41].

Component Theorem. Let G be a group with $O(G) = 1$ satisfying the B-conjecture. Then, with known exceptions, the members of $\mathcal{L}^*$ are standard subgroups of G.

A quasisimple subgroup L of G is a <u>standard subgroup</u> of G if $K = C_G(L)$ is tightly embedded in G, $N_G(K) = N_G(L)$, and L commutes with none of its conjugates.

A subgroup K of G is <u>tightly embedded</u> in G if K has even order while the intersections of K with any distinct conjugate is of odd order.

If x is an involution in G centralizing a standard subgroup L then $C(x)$ is said to be in standard form. The requirement that $C(L)$ be tightly embedded in G seems to be the right restriction on the size of $C(L)$. Counter examples tell us we cannot hope to prove $C(L)$ has cyclic Sylow 2-groups in general. The concept of tight embeddings seems to be important in finite group theory. It also figures prominently in the analysis of groups of characteristic 2-type.

Evidentially, the relation $\longrightarrow$ is suggested by L-balance. For if $L \longrightarrow K$ with L a component of $C(x)$, K a component of $C(y)$, and $[x,y] = 1$, then L-balance tells us either $L \leq K$ or $L \leq KK^x$

with L a homomorphic image of K. Let $L \in L^*$ and assume for simplicity, that the second type of embedding never occurs. Then by L-balance, L is a component of C(y) for each involution y centralizing L. If we assume further that L commutes with none of its conjugates, then it is trivial to check that L is standard in G. While all difficulties are concealed by the two assumptions, this nevertheless gives a general idea of how to prove the Component Theorem

This brings us to the standard form problems. Let X be a known simple group. The <u>standard form problem</u> for X is to show that if H is a simple group, $H \leq G \leq \text{Aut}(H)$, and L is a standard subgroup of G with $L/Z(L) \cong X$, then H is a known simple group. A solution of the standard form problem for X is a verification of the fundamental principle for centralizers of involutions in standard form with component of type X.

The standard form problems for most simple groups have been solved; perhaps a dozen of the problems remain to be completed. These last few standard form problems are the only remaining obstacle to establishing the Component Group Conjecture. Presumably, the standard form problems that remain are no more difficult than many which are already complete.

The present state of the art in the theory of groups of component type may be summarized as follows:

Theorem 5.3. Let G be a minimal counter example to the Component Group Conjecture. Then G is balanced and contains a standard subgroup L such that C(L) has cyclic Sylow 2-subgroups and $L/Z(L)$ is a group of Lie type of small Lie rank over a field of order at most 4, $^2F_4(2^n)$, or one of several sporadic groups.

Groups of Component Type / 47

In the last two chapters, I have attempted to present a broad outline of the proof of this result. If and when the Component Group Conjecture is established, the classification of the finite simple groups is reduced to a classification of the groups of characteristic 2-type. For if G is a simple group minimal subject to not appearing on the list of known simple groups, then by the Unbalanced Group Conjecture and the Component Group Conjecture, $O(C(x)) = E(C(x)) = 1$ for each involution x in G. It is now an easy exercise to show G is of characteristic 2-type. The next chapter considers the problem of classifying the groups of characteristic 2-type.

VI. Groups of Even Characteristic

We now come to the groups of characteristic 2-type. This is the area where the majority of the work to classify the finite simple groups remains. However, the general outline of an approach to classifying the groups of characteristic 2-type is available, many steps in that outline are complete, and work is in progress to finish the remaining steps. This chapter discusses that outline.

Most of the known simple groups of characteristic 2-type are groups of Lie type and even characteristic. The analysis described in this chapter reflects that fact. Following the discussion in earlier chapters, we attempt to show there is an element of odd prime order whose centralizer is in standard form. We try to pick the odd prime p to be as involved as possible in the 2-local structure of G. To do this, we pick p so that the p-rank of some 2-local is large. This leads to the following concept.

The <u>2-local p-rank</u> of G is

$$m_{2,p}(G) = \max \{m_p(M): M \text{ 2-local in } G\}$$

We are also interested in a second parameter:

$$e(G) = \max \{m_{2,p}(G): p \text{ odd prime}\}$$

If G is of Lie type and even characteristic, then $e(G)$ approximates the Lie rank of G. This rank seems to the significant measure of size for the groups of characteristic 2-type. The groups with $e(G) \leq 2$ are our "small" groups and require special attention and techniques. G is said to be <u>quasithin</u> if $e(G) \leq 2$. G is <u>thin</u> if $e(G) = 1$. Hence G is thin precisely when each 2-local subgroup of G has cyclic Sylow p-subgroups for each odd prime p.

The thin groups were classified by the author in [9] extending earlier work of Janko [26] and Thompson [36]. G. Mason has recently announced significant progress toward a classification of the quasithin groups.

The methods used by all these authors are similar in spirit to the methods used in one of the cases in the generic analysis and will be discussed later.

Let us turn then to the generic situation where $e(G) \geq 3$. Let K be the set of all known simple groups. A group G is a K-group if each composition factor of each subgroup of G is in K. We wish to show K is the set of all simple groups. A minimal counter example to such a theorem will have all proper subgroups K-groups. In the generic characteristic 2-type analysis, we assume this minimal counter example G to be of characteristic 2-type with $e(G) \geq 3$.

The following theorem is the cornerstone of the generic characteristic 2-type analysis:

Theorem 6.1. Let G be simple of characteristic 2-type with $e(G) \geq 3$ and with all proper subgroups K-groups. Then one of the following hold:

(1) G possesses elements of odd prime order whose centralizer is in standard form.

(2) G possesses a 2-local M such that $O_2(M)$ has no noncyclic characteristic abelian subgroups.

(3) For many odd primes p, G has an almost strongly p-embedded 2-local subgroup.

The concept of an "almost strongly p-embedded subgroup" is rather technical. For the purposes of the exposition here, we shall assume such subgroups are actually strongly p-embedded as defined in Chapter IV.

The proof of Theorem 6.1 is not yet in final form. The case $e(G) > 3$ is due to Gorenstein and Lyons; the case $e(G) = 3$ is due to the author. The theorem is proved using signalizer functor analysis. This is one of the reasons for the partition into groups with $e(G) \geq 3$ and $e(G) < 3$. In the latter case, there are no elementary abelian p-subgroups of rank 3 in 2-locals to which we can associate signalizer functors.

The fact that G is of characteristic 2-type is used to great advantage and makes certain parts of the analysis easier than corresponding sections of the analysis of groups of component type. Let x be an element of odd prime order p which is a candidate for an element whose centralizer H is in standard form. We choose x to lie in an elementary abelian p-subgroup A of rank $m_{2,p}(G)$ which is contained in a 2-local of G. Recall that the purpose of the signalizer functor analysis is to show $O_{p'}(H)$ small. As we know G to be of characteristic 2-type, this essentially amounts to showing $O_2(H)$ small. An A-signalizer functor θ is defined. If θ(A) = 1, then H may usually be chosen to be in standard form. Indeed, we have much more. As G is of characteristic 2-type, either E(H) is of Lie type and even characteristic or the exceptional second case of Theorem 6.1 arises.

Moreover, in the first case the embedding of A in AE(H) is highly determined: $A \cap E(H)$ is essentially a maximal torus of E(H) and $<x> = C_A(E(H))$. As indicated in Chapter IV, the case where $\theta(A) \neq 1$ leads to a strongly p-embedded subgroup. As G is of characteristic 2-type, this subgroup is even a 2-local. This is how the third case of Theorem 6.1 arises.

In the remainder of this chapter, we consider each of the three cases arising out of Theorem 6.1. The fundamental principle suggests the first case can be done. The author has shown, for example, that if case one holds with $e(G) = 3$, then either $G \in K$ or the prime in question is 3 and the component of the centralizer in standard form is $L_4(2)$, $L_5(2)$, of $Sp_6(2)$.

Let us now consider the second case of Theorem 6.1. Recall our example of the general linear group. Consider $G = GL_n(2)$ and let u be the unipotent element in Figure 2 in Chapter II with $r = 1$. That is, u induces a transaction on the n-dimensional vector space over GF(2) preserved by G. Then $C_G(u)$ is a 2-local subgroup of G such that the group $U = O_2(C(u))$ of Figure 2 has no noncyclic characteristic abelian subgroups. As a matter of fact, most groups of Lie type over GF(2) possess such a 2-local. Conversely, all but a finite number of simple groups with such a 2-local are of Lie type over GF(2) as the following theorem indicates:

Theorem 6.2. Let G be a finite simple group possessing a 2-local M such that $C_M(O_2(M)) \leq O_2(M)$ and $O_2(M)$ has no noncyclic characteristic abelian subgroups. Then G is a group of Lie type over the field of order 2, $G_2(3)$, $U_4(3)$, $L_4(3)$, $\Omega_8^+(3)$, A_9, $L_2(2^n \pm 1)$, M_{11}, M_{12}, M_{24}, HS, He, Sz, HJ, J_3, J_4, Co_2, Co_1, F_{24}, F_5, F_3, F_2, or F_1.

Many group theorists have contributed to the proof of Theorem 6.2. Possibly the largest contribution is by Timmesfeld in [38]. Other contributions were made by the author [6,7], Dempwolf and S. Wong [13], Gorenstein and Harada [23], F. Smith [29, 30], S. Smith [31, 32, 33], and Thompson.

I suggested that in Chapter III that it was usually difficult to use the centralizer of an element x of order p with $C(O_p(C(x))) \leq O_p(C(x))$ to determine a simple group. The situation in Theorem 6.2 is exceptional because the hypothesis that $O_2(M)$ have no noncyclic characteristic abelian subgroups precisely determines $O_2(M)$ by a theorem of P. Hall. In the most interesting case, $O_2(M) = Q$ is **extraspecial**. That is $Z(Q) = [Q,Q]$ is of order 2 and $\tilde{Q} = Q/Z(Q)$ is elementary abelian. Now an easy construction makes $\tilde{Q}$ into an orthogonal space over $GF(2)$ upon which M/Q is faithfully represented. This makes it possible to determine M and then G. The Fischer theory is quite useful in this connection.

In any case, Theorem 6.2 supplies a definitive answer as to which groups arise in the second case of Theorem 6.1. There is another consequence of Theorem 6.2 worth mentioning. I have listed the groups satisfying the hypothesis of Theorem 6.2 to make a point: observe that 16 of the 26 sporadic groups appear on this list. With the exception of M_{22} and M_{23} all other aporadic groups are of component type, and the two Mathieu groups may be regarded as small exceptions. Hence, Theorems 5.3 and 6.2 may be regarded as heuristic evidence that few if any, sporadic groups remain to be discovered.

This brings us to the third case of Theorem 6.1. I refer to this case as the <u>Uniqueness Case</u>. The effect of the hypothesis is to show that for a large set $\sigma(G)$ of odd primes, for each $p \in \sigma(G)$, and for each noncyclic p-subgroup X of G, X is contained in a unique maximal 2-local M of G. Here we partially order the set of 2-local subgroups of G and define the set $\mathcal{M}$ of maximal 2-locals to be the maximal members of this partially ordered set.

The Uniqueness Case first appears in Thompson's classification of N-groups [36]. An N-group is a simple group in which all local subgroups are solvable. Many of the concepts in this chapter and indeed many of the techniques in the classification program originate in the N-group paper and the Odd Order paper. At least half of the N-group paper is devoted to the Uniqueness Case and to the quasithin case. The methods used in both cases are similar. They are the same methods used in the general Uniqueness Case and general quasithin analysis.

In all cases, we start with a set Ω of <u>uniqueness subgroups</u>. That is, each member of Ω is contained in a unique member of $\mathcal{M}$. In the Uniqueness Case and in Thompson's analysis of N-groups with $e(G) = 2$, Ω is a set of noncyclic p-groups, for certain odd primes p. In the generic quasithin analysis Ω consists of certain minimal non-solvable subgroups of G.

Let $H \leq G$ with $C(O_2(H)) \leq O_2(H)$ and let T be a Sylow 2-group of H. We attempt to produce <u>factorizations</u> of H of the form

$$H = C_H(C)N_H(W) \quad \text{or,}$$

$$H = <C_H(C), N_H(W)>$$

for certain cannonically defined subgroups C and W of T. For example, we hope to produce distinct members M and H of $\mathcal{M}$ containing a Sylow 2-group T of G. From the factorization $M = C_M(C)N_M(W)$ we hope to produce X and Y in Ω contained in $C_M(C)$ and $N_M(W)$, respectively. As X and Y are uniqueness subgroups, this means $C_G(C)$ and $N_G(W)$ are contained in M. But then $H = C_H(C)N_H(W)$ is contained in M, contradicting the maximality of H.

Such analysis tends to be highly intricate and as a result appears to be understood by relatively few group theorists and well understood by no one. While both the Uniqueness Case and the quasithin groups will presumably be dealt with within the next few years, it will probably take much longer to produce the definitive treatment of these problems.

Just as the prime 2 plays a special role in the analysis of finite simple groups, the prime 3 plays a special role in the analysis of groups of characteristic 2-type and most particularly in the Uniqueness Case. In particular, in the generic situation G has a strongly 3-embedded 2-local of high 3-rank. The author has shown that no groups G satisfy the uniqueness case with $m_{2,3}(G) > 3$, hence handling the generic case. Significant progress has been made toward extending this work to the general Uniqueness Case.

VII. The Future

Let us recall the steps in the classification program which remain to be completed. We have seen that the classification is divided into the classification of the groups of component type and of characteristic 2-type. A classification of the groups of component type amounts to a verification of the Component Group Conjecture appearing in Chapter V. Theorem 5.3 summarizes the current progress toward verification of the conjecture. In particular, we see that the conjecture is reduced to a solution of one of a small number of standard form problems. Such problems are discussed briefly in Chapter V.

Work remaining in the classification of the groups of characteristic 2-type is more substantial. There are three major problems to be completed here. First, the quasithin groups must be classified. This brings us to the generic situation where Theorem 6.1 is applicable and where a classification of the groups satisfying the conclusion of case one or case three of Theorem 6.1 remains to be completed. In all three areas, there exist substantial partial results. In addition to these three major problems there are several more specialized problems which need to be solved to complete the general theory of groups of characteristic 2-type, used in the analysis of the major problems. At this point, it might be helpful to refer again to the flow diagram at the end of Chapter II.

Frequently of late, I am asked: "What will you finite group theorists do when the simple groups are classified?" In the first place, it is evident that if and when a classification is achieved, the proof will require a great deal of attention and modification before it can be regarded as definitive. Actually some such modification is already going on before the fact of the classification. H. Bender, for example, has produced a number of beautiful simplifications of the proofs of major results. It is interesting to note however, that these modifications all take place within the framework of the existing classification program and utilize existing techniques. Thus my first answer to the question raised above would be that even after a first proof is achieved much of the attention of finite simple group theorists will still be directed toward the classification. A problem of this magnitude is not solved abruptly, rather it will take a number of years before even the specialists come to really understand the solution and many more before it is modified and simplified to be accessible to other mathematicians.

In the second place, it is the case that for at least two decades the classification has received the majority of attention of many finite group theorists. In the process, the mathematical interests of many of us have become quite specialized. If and when a classification is achieved, there will be time to take stock of the situation and think about the possibility of applying the techniques of finite simple group theory, and indeed the classification itself, to other problems.

Several such problems suggest themselves already. The representation of the simple groups as matrix groups is certainly an area of great interest, and has been since the nineteenth century.

The representations of the groups of Lie type are of particular interest. Another question which has received some attention is the determination of the primitive permutation representations of the simple groups. This amounts to determining the maximal subgroups of the simple groups. Almost certainly a classification would be of much assistance in this problem. There has also been a certain amount of interaction between finite group theory and the theory of finite geometries. Probably there would be more after a classification.

These are three existing problem areas. It is perhaps more exciting to speculate that some as yet unsuspected problems will emerge after the classification to command the attention of finite group theorists.

References

1. M. Aschbacher, Groups generated by odd transpositions, I, Math. Z. 127 (1972), 45-56, II, III, and IV, J. Alg. 26 (1973), 451-491.

2. _____, Finite groups with a proper 2-generated core, Trans. AMS. 197 (1974), 87-112.

3. _____, F-sets and permutation groups, J. Alg. 30 (1974), 400-416.

4. _____, On finite groups of component type, Illinois J. Math. 19 (1975), 87-115.

5. _____, A homomorphism theorem for finite graphs, Proc. AMS. 54 (1976), 468-470.

6. _____, On finite groups in which the generalized Fitting group of the centralizer of some involution is symplectic but not extraspecial, Comm. Alg. 4 (1976), 595-616.

7. _____, On finite groups in which the generalized Fitting group of the centralizer of some involution is extraspecial, Illinois J. Math. 21 (1977), 347-364.

8. _____, A characterization of Chevalley groups over fields of odd order, Ann. Math. 106 (1977), 353-468.

9. _____, Thin finite simple groups, J. Alg. 54 (1978), 50-152.

10. M. Aschbacher and M. Hall, Jr., Groups generated by a class of elements of order 3, J. Alg. 24 (1973), 591-612.

11. H. Bender, Transitive Gruppen gerader Ordnung, in denen jede Involution genau einer Punkt festlasst, J. Alg. 17 (1971), 527-554.

12. R. Carter, Simple Groups of Lie Type, Wiley, New York, 1972.

13. U. Dempwolf and S. Wong, On finite groups whose centralizer of an involution has normal extraspecial and abelian subgroups, I, II, J. Alg. 45 (1977), 247-253, 52 (1978), 210-217.

14. B. Fischer, Finite groups generated by 3-transpositions, I, Inventiones Math. 13 (1971), 232-246.

15. P. Fong and G. Seitz, Groups with a (B,N)-pair of rank 2, I, II, Inventiones Math. 21 (1973), 1-57, 24(1974), 191-239.

16. G. Glauberman, Central elements of core-free groups, J. Alg. 4 (1966), 403-420.

17. _____, On solvable signalizer functors in finite groups, J. London Math. Soc. 33 (1976), 1-27.

18. D. Goldschmidt, Solvable signalizer functors in finite groups, J. Alg. 21 (1972), 137-148.

19. _____, 2-signalizer functors on finite groups, J. Alg. 21 (1972), 321-340.

20. D. Gorenstein, Finite Groups, Harper and Row, New York, 1968.

21. _____, On the centralizers of involutions in finite groups, J. Alg. 11 (1969), 243-277.

22. _____, The flatness of signalizer functors on finite groups, J. Alg. 13 (1969), 509-512.

23. D. Gorenstein and K. Harada, Finite groups whose 2-subgroups are generated by at most 4 elements, Memoirs AMS. 147 (1974), 1-464.

24. D. Gorenstein and J. Walter, Balance and generation in finite groups, J. Alg. 33 (1975), 224-287.

25. C. Hering and W. Kantor, and G. Seitz, Finite groups with a split (B,N)-pair of rank 1, J. Alg. 20 (1972), 435-475.

26. Z. Janko, Nonsolvable finite groups all of whose 2-local subgroups are solvable, I, J. Alg. 21 (1972), 458-517.

27. A. MacWilliams, On 2-groups with no normal abelian subgroups of rank 3 and their occurrence as Sylow 2-subgroups of finite simple groups, Trans. AMS. 150 (1970), 345-408.

28. M. Powell and G. Higman, ed., Oxford Conference on Finite Simple Groups, Academic Press, London, 1969.

29. F. Smith, On a result of M. Aschbacher, Comm. Alg. 51(1977) 279-288.

30. _____, On finite groups with large extraspecial 2-subgroups, J. Alg. 44 (1977), 477-487.

31. S. Smith, A characterization of orthogonal groups over GF(2), to appear in J. Alg.

32. _____, A characterization of finite Chevalley and twisted groups of type E over GF(2), to appear in J. Alg.

33. _____, Large extraspecial subgroups of widths 4 and 6, to appear in J. Alg.

34. B. Stark, Another look at Thompson's quadratic pairs, J. Alg. 45 (1977), 334-342.

35. M. Suzuki, Finite groups of even order in which Sylow 2-subgroups are independent, Ann. Math. 80 (1964), 58-77.

36. J. Thompson, Nonsolvable finite groups all of whose local subgroups are solvable, I, Bull. AMS. 74 (1968), 383-437, II-VI, Pacific J. Math. 33 (1970), 451-536, 39 (1971), 483-534, 48 (1973), 511-592, 50 (1974), 215-297, 51 (1974), 573-630.

37. F. Timmesfeld, Groups generated by root involutions, I, II, J. Alg. 33 (1975), 75-134, 35 (1975), 367-441.

38. _____, Finite simple groups in which the generalized Fitting group of the centralizer of some involution is extraspecial, Ann. Math. 107 (1978), 297-369.

39. J. Tits, Buildings of spherical type and Finite BN-pairs, Springer-Verlag, Berlin, 1974.

40. E. Witt, Die 5-fach transitiveri Gruppen von Mathieu, Abh. Math. Sem. Univ. Hamburg 121 (1938), 256-264.

41. R. Foote, Finite groups with components of 2-rank 1, I, II, J. Alg. 41 (1976), 16-57.